JOHN HERSCHEL

GREAT ASTRONOMERS

ROBERT STAWELL BALL

JOHN HERSCHEL

GREAT ASTRONOMERS

BY

ROBERT STAWELL BALL

D.Sc, LL.D., F.R.S.

(July 1, 1840 – November 25, 1913)

LOWNDEAN PROFESSOR OF ASTRONOMY AND GEOMETRY IN THE UNIVERSITY OF CAMBRIDGE

DIAMOND™ BOOKS

www.diamondbooks.ca

TORONTO, CANADA – 2017

DIAMOND BOOKS - CANADA

Toronto, ON, CANADA
http://www.diamondbooks.ca

BIBLIOGRAPHIC INFORMATION

'JOHN HERSCHEL'
Is an extract or a chapter from the book
"GREAT ASTRONOMERS" by Robert Stawell Ball.
It was first published in 1896.

PUBLISHED IN CANADA

Published in Canada by DIAMOND BOOKS - CANADA, an imprint of DIAMOND PUBLISHERS - http://www.diamondpublishers.com

First Edition: April, 2017.

PAPERBACK EDITION : ISBN: 978-1-988357-64-5
E-BOOK EDITION : ISBN: 978-1-988357-65-2

PRINTED IN CANADA

PREFACE
TO THE ORIGINAL EDITION

It has been my object in these pages to present the life of each astronomer in such detail as to enable the reader to realise in some degree the man's character and surroundings; and I have endeavoured to indicate as clearly as circumstances would permit the main features of the discoveries by which he has become known.

There are many types of astronomers—from the stargazer who merely watches the heavens, to the abstract mathematician who merely works at his desk; it has, consequently, been necessary in the case of some lives to adopt a very different treatment from that which seemed suitable for others.

ROBERT STAWELL BALL.

The Observatory, Cambridge.
October, 1895.

NOTE: The full-length of 'Preface' can be found in its original edition of the book "GREAT ASTRONOMERS".

INTRODUCTION

Of all the natural sciences there is not one which offers such sublime objects to the attention of the inquirer as does the science of astronomy. From the earliest ages the study of the stars has exercised the same fascination as it possesses at the present day. Among the most primitive peoples, the movements of the sun, the moon, and the stars commanded attention from their supposed influence on human affairs. The practical utilities of astronomy were also obvious in primeval times. Maxims of extreme antiquity show how the avocations of the husbandman are to be guided by the movements of the heavenly bodies. The positions of the stars indicated the time to plough, and the time to sow. To the mariner who was seeking a way across the trackless ocean, the heavenly bodies .offered the only reliable marks by which his path could be guided. There was, accordingly, a stimulus both from intellectual curiosity and from practical necessity to follow the movements of the stars. Thus began a search for the causes of the ever-varying phenomena which the heavens display.

- ROBERT STAWELL BALL

Many of the earliest discoveries are indeed prehistoric. The great diurnal movement of the heavens, and the annual revolution of the sun, seem to have been known in times far more ancient than those to which any human monuments can be referred. The acuteness of the early observers enabled them to single out the more important of the wanderers which we now call planets. They saw that the star-like objects, Jupiter, Saturn, and Mars, with the more conspicuous Venus, constituted a class of bodies wholly distinct from the fixed stars among which their movements lay, and to which they bear such a superficial resemblance. But the penetration of the early astronomers went even further, for they recognized that Mercury also belongs to the same group, though this particular object is seen so rarely. It would seem that eclipses and other phenomena were observed at Babylon from a very remote period, while the most ancient records of celestial observations that we possess are to be found in the Chinese annals. The study of astronomy, in the sense in which we understand the word, may be said to have commenced under the reign of the Ptolemies at Alexandria. The most famous name in the science of this period is that of Hipparchus, who lived and worked at Rhodes about the year 160 B.C. It was his splendid investigations that first wrought the observed facts into a coherent branch of knowledge. He recognized the primary obligation which lies on the student of the heavens to compile as complete an inventory as possible of the objects which are there to be found. Hipparchus accordingly commenced by undertaking, on a small scale, a task exactly similar to that on which modern astronomers, with all available appliances of meridian circles, and photographic telescopes, are constantly engaged at the present day. He compiled a catalogue of the principal fixed stars, which is of special value to astronomers, as being the earliest work of its kind which has been handed down. He also studied the movements of the sun and the moon, and framed theories to account for the incessant changes which he saw in progress. He found a much more difficult problem in his attempt to interpret satisfactorily the complicated movements of the planets. With the view of constructing a theory which should give some coherent account of the subject, he made many observations of the places of these wandering stars. How great were the advances which Hipparchus accomplished may be appreciated if we reflect that, as a preliminary task to his more purely astronomical labours, he had to invent that branch of mathematical science by which alone the problems he proposed could be solved. It was for this purpose that he devised the indispensable method of calculation which we now know so well as trigonometry. Without the aid rendered by this beautiful art it would have been impossible for any n ally important advance in astronomical calculation to have been effected.

But the discovery which shows, beyond all others, that Hipparchus possessed one of the master-minds of all time was the detection of that remarkable celestial movement known as the precession of the equinoxes. The inquiry which conducted to this discovery involved a most profound investigation, especially when it is remembered that in the days of Hipparchus the means of observation of the heavenly bodies were only of the rudest description, and the available observations of earlier dates were extremely scanty. We can but look with astonishment on the genius of the man who, in spite of such difficulties, was able to detect such a phenomenon as the precession, and to exhibit its actual magnitude. I shall endeavour to explain the nature of this singular celestial movement, for it may be said to offer the first instance in the history of science in which we find that combination of accurate observation with skilful interpretation, of which, in the subsequent development of astronomy, we have so many splendid examples.

The word equinox implies the condition that the night is equal to the day. To a resident on the equator the night is no doubt equal to the day at all times in the year, but to one who lives on any other part of the earth, in either hemisphere, the night and the day are not generally equal. There is, however, one occasion in spring, and another in autumn, on which the day and the night are each twelve hours at all places on the earth. When the night and day are equal in spring, the point which the sun occupies on. the heavens is termed the vernal equinox. There is similarly another point in which the sun is situated at the time of the autumnal equinox. In any investigation of the celestial movements the positions of these two equinoxes on the heavens are of primary importance, and Hipparchus, with the instinct of genius, perceived their significance, and commenced to study them. It will be understood that we can always define the position of a point on the sky with reference to the surrounding stars. No doubt we do not see the stars near the sun when the sun is shining, but they are there nevertheless. The ingenuity of Hipparchus enabled him to determine the positions of each of the two equinoxes relatively to the stars which lie in its immediate vicinity. After examination of the celestial places of these points at different periods, he was led to the conclusion that each equinox was moving relatively to the stars, though that movement was so slow that twenty five thousand years would necessarily elapse before a complete circuit of the heavens was accomplished. Hipparchus traced out this phenomenon, and established it on an impregnable basis, so that all astronomers have ever since recognised the precession of the equinoxes) as one of the fundamental facts of astronomy. Not until nearly two thousand years after Hipparchus had made this splendid discovery was the explanation of its cause given by Newton.

From the days of Hipparchus down to the present hour the science of astronomy has steadily grown. One great observer after another has appeared from time to time, to reveal some new phenomenon with regard to the celestial bodies or their movements, while from time to time one commanding intellect after another has arisen to explain the true import of the facts of observations. The history of astronomy thus becomes inseparable from the history of the great men to whose labours its development is due. In the ensuing chapters we have endeavoured to sketch the lives and the work of the great philosophers, by whose labours the science of astronomy has been created.

We shall commence with Ptolemy, who, after the foundations of the science had been laid by Hipparchus, gave to astronomy the form in which it was taught throughout the Middle Ages. We shall next see the mighty revolution in our conceptions of the universe which are associated with the name of Copernicus. We then pass to those periods illumined by the genius of Galileo and Newton, and afterwards we shall trace the careers of other more recent discoverers, by whose industry and genius the boundaries of human knowledge have been so greatly extended. Our history will be brought down late enough to include some of the illustrious astronomers who laboured in the generation which has just passed away.

- ROBERT STAWELL BALL

TABLE OF CONTENTS

CHAPTERS

**An original Chapter from the book: "GREAT ASTRONOMERS", by Robert Stawell Ball.*

CHAPTER 13

GREAT ASTRONOMERS

John Herschel

(March 7, 1792 – May 11, 1871)

John Frederick William **Herschel**
(March 7, 1792 – May 11, 1871)

CHAPTER 13

GREAT ASTRONOMERS: JOHN HERSCHEL

THIS illustrious son of an illustrious father was born at Slough, near Windsor, on the 7th March, 1792. He was the only child of Sir William Herschel, who had married somewhat late in life, as we have already mentioned.

The surroundings among which the young astronomer was reared afforded him an excellent training for that career on which he was to enter, and in which he was destined to attain a fame only less brilliant than that of his father. The circumstances of his youth permitted him to enjoy one great advantage which was denied to the elder Herschel. He was able, from his childhood, to devote himself almost exclusively to intellectual pursuits. William Herschel, in the early part of his career, had only been able to snatch occasional hours for study from his busy life as a professional musician. But the son, having been born with a taste for the student's life, was fortunate enough to have been endowed with the leisure and the means to enjoy it from the commencement. His early years have been so well described by the late Professor Pritchard in the "Report of the Council of the Royal Astronomical Society for 1872," that I venture to make an extract here:—

"A few traits of John Herschel's boyhood, mentioned by himself in his maturer life, have been treasured up by those who were dear to him, and the record of some of them may satisfy a curiosity as pardonable as inevitable, which craves to learn through what early steps great men or great nations become illustrious. His home was singular, and singularly calculated to nurture into greatness any child born as John Herschel was with natural gifts, capable of wide development. At the head of the house there was the aged, observant, reticent philosopher, and rarely far away his devoted sister, Caroline Herschel, whose labours and whose fame are still cognisable as a beneficent satellite to the brighter light of her illustrious brother.

It was in the companionship of these remarkable persons, and under the shadow of his father's wonderful telescope, that John Herschel passed his boyish years. He saw them, in silent but ceaseless industry, busied about things which had no apparent concern with the world outside the walls of that well-known house, but which, at a later period of his life, he, with an unrivalled eloquence, taught his countrymen to appreciate as foremost among those living influences which but satisfy and elevate the noblest instincts of our nature. What sort of intercourse passed between the father and the boy may be gathered from an incident or two which he narrated as having impressed themselves permanently on the memory of his youth. He once asked his father what he thought was the oldest of all things.

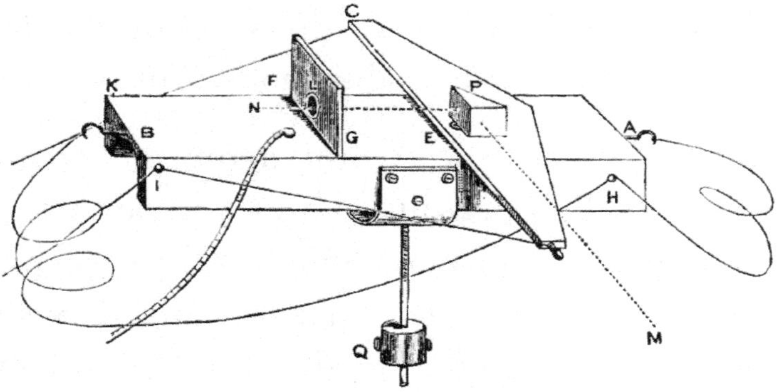

Astronometer made by **Sir J. Herschel** to compare the light of certain stars by the intervention of the moon.

The father replied, after the Socratic method, by putting another question: 'And what do you yourself suppose is the oldest of all things?' The boy was not successful in his answers, thereon the old astronomer took up a small stone from the garden walk: 'There, my child, there is the oldest of all the things that I certainly know.' On another occasion his father is said to have asked the boy, 'What sort of things, do you think, are most alike?' The delicate, blue-eyed boy, after a short pause, replied, 'The leaves of the same tree are most like each other.' 'Gather, then, a handful of leaves of that tree,' rejoined the philosopher, 'and choose two that are alike.' The boy failed; but he hid the lesson in his heart, and his thoughts were revealed after many days. These incidents may be trifles; nor should we record them here had not John Herschel himself, though singularly reticent about his personal emotions, recorded them as having made a strong impression on his mind Beyond all doubt we can trace therein, first, that grasp and grouping of many things in one, implied in the stone as the oldest of things; and, secondly, that fine and subtle discrimination of each thing out of many like things as forming the main features which characterized the habit of our venerated friend's philosophy."

John Herschel entered St. John's College, Cambridge, when he was seventeen years of age. His university career abundantly fulfilled his father's eager desire, that his only son should develop a capacity for the pursuit of science. After obtaining many lesser distinctions, he finally came out as Senior Wrangler in 1813. It was, indeed, a notable year in the mathematical annals of the University. Second on that list, in which Herschel's name was first, appeared that of the illustrious Peacock, afterwards Dean of Ely, who remained throughout life one of Herschel's most intimate friends.

Almost immediately after taking his degree, Herschel gave evidence of possessing a special aptitude for original scientific investigation. He sent to the Royal Society a mathematical paper which was published in the Philosophical Transactions. Doubtless the splendour that attached to the name he bore assisted him in procuring an early recognition of his own great powers. Certain it is that he was made a Fellow of the Royal Society at the unprecedentedly early age of twenty-one. Even after this remarkable encouragement to adopt a scientific career as the business of his life, it does not seem that John Herschel at first contemplated devoting himself exclusively to science. He commenced to prepare for the profession of the Law by entering as a student at the Middle Temple, and reading with a practising barrister.

But a lawyer John Herschel was not destined to become. Circumstances brought him into association with some leading scientific men. He presently discovered that his inclinations tended more and more in the direction of purely scientific pursuits. Thus it came to pass that the original intention as to the calling which he should follow was gradually abandoned. Fortunately for science

Herschel found its pursuit so attractive that he was led, as his father had been before him, to give up his whole life to the advancement of knowledge. Nor was it unnatural that a Senior Wrangler, who had once tasted the delights of mathematical research, should have been tempted to devote much time to this fascinating pursuit. By the time John Herschel was twenty-nine he had published so much mathematical work, and his researches were considered to possess so much merit, that the Royal Society awarded him the Copley Medal, which was the highest distinction it was capable of conferring.

At the death of his father in 1822, John Herschel, with his tastes already formed for a scientific career, found himself in the possession of ample means. To him also passed all his father's great telescopes and apparatus. These material aids, together with a dutiful sense of filial obligation, decided him to make practical astronomy the main work of his life. He decided to continue to its completion that great survey of the heavens which had already been inaugurated, and, indeed, to a large extent accomplished, by his father.

The first systematic piece of practical astronomical work which John Herschel undertook was connected with the measurement of what are known as "Double Stars." It should be observed, that there are in the heavens a number of instances in which two stars are seen in very close association. In the case of those objects to which the expression "Double Stars" is generally applied, the two luminous points are so close together that even though they might each be quite bright enough to be visible to the unaided eye, yet their proximity is such that they cannot be distinguished as two separate objects without optical aid. The two stars seem fused together into one. In the telescope, however, the bodies may be discerned separately, though they are frequently so close together that it taxes the utmost power of the instrument to indicate the division between them.

The appearance presented by a double star might arise from the circumstance that the two stars, though really separated from each other by prodigious distances, happened to lie nearly in the same line of vision, as seen from our point of view. No doubt, many of the so-called double stars could be accounted for on this supposition. Indeed, in the early days when but few double stars were known, and when telescopes were not powerful enough to exhibit the numerous close doubles which have since been brought to light, there seems to have been a tendency to regard all double stars as merely such perspective effects. It was not at first suggested that there could be any physical connection between the components of each pair. The appearance presented was regarded as merely due to the circumstance that the line joining the two bodies happened to pass near the earth.

In the early part of his career, Sir William Herschel seems to have entertained the view then generally held by other astronomers with regard to the nature of these stellar pairs. The great observer thought that the double stars could therefore be made to afford a means of solving that problem in which so many of the observers of the skies had been engaged, namely, the determination of the distances of the stars from the earth. Herschel saw that the displacement of the earth in its annual movement round the sun would produce an apparent shift in the place of the nearer of the two stars relatively to the other, supposed to be much more remote. If this shift could be measured, then the distance of the nearer of the stars could be estimated with some degree of precision.

As has not unfrequently happened in the history of science, an effect was perceived of a very different nature from that which had been anticipated. If the relative places of the two stars had been apparently deranged merely in consequence of the motion of the earth, then the phenomenon would be an annual one. After the lapse of a year the two stars would have regained their original relative positions. This was the effect for which William Herschel was looking. In certain of the so-called double stars, he, no doubt, did find a movement. He detected the remarkable fact that both the apparent distance and the relative positions of the two bodies were changing. But what was his surprise to observe that these alterations were not of an annually periodic character. It became evident then that in some cases one of the component stars was actually revolving round the other, in an orbit which required many years for its completion. Here was indeed a remarkable discovery. It was clearly impossible to suppose that movements of this kind could be mere apparent displacements, arising from the annual shift in our point of view, in consequence of the revolution of the earth. Herschel's discovery established the interesting fact that, in certain of these double stars, or binary stars, as these particular objects are more expressively designated, there is an actual orbital revolution of a character similar to that which the earth performs around the sun. Thus it was demonstrated that in these particular double stars the nearness of the two components was not merely apparent. The objects must actually lie close together at a distance which is small in comparison with the distance at which either of them is separated from the earth. The fact that the heavens contain pairs of twin suns in mutual revolution was thus brought to light.

In consequence of this beautiful discovery, the attention of astronomers was directed to the subject of double stars with a degree of interest which these objects had never before excited. It was therefore not unnatural that John Herschel should have been attracted to this branch of astronomical work. Admiration for his father's discovery alone might have suggested that the son should strive to develop this territory newly opened up to research. But it also happened that the mathematical talents of the younger Herschel inclined his inquiries in the same direction. He saw clearly that, when sufficient observations of any particular binary star had been accumulated, it would then be within the power of the mathematician to elicit from those observations the shape and the position in space of the path which each of the revolving stars described around the other.

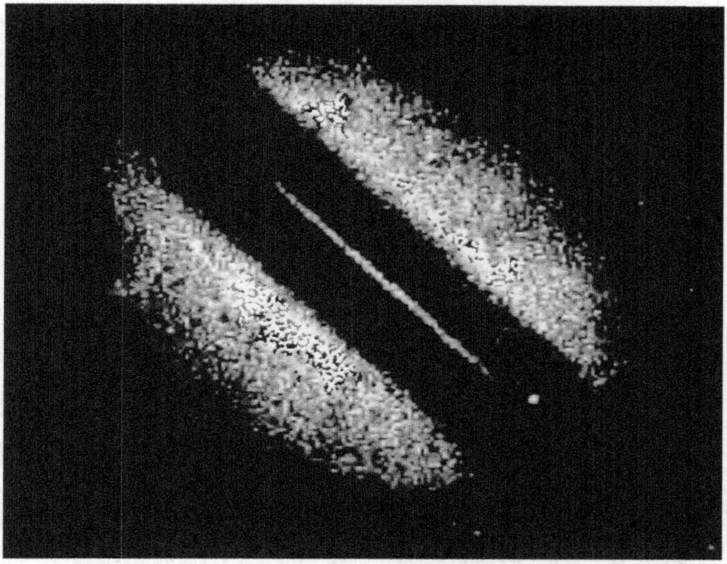

Nebula in Southern Hemisphere, drawn by Sir John Herschel.

Indeed, in some cases he would be able to perform the astonishing feat of determining from his calculations the weight of these distant suns, and thus be enabled to compare them with the mass of our own sun. But this work must follow the observations, it could not precede them The first step was therefore to observe and to measure with the utmost care the positions and distances of those particular double stars which appear to offer the greatest promise in this particular research. In 1821, Herschel and a friend of his, Mr. James South, agreed to work together with this object. South was a medical man with an ardent devotion to science, and possessed of considerable wealth. He procured the best astronomical instruments that money could obtain, and became a most enthusiastic astronomer and a practical observer of tremendous energy.

South and John Herschel worked together for two years in the observation and measurement of the double stars discovered by Sir William Herschel. In the course of this time their assiduity was rewarded by the accumulation of so great a mass of careful measurements that when published, they formed quite a volume in the Philosophical Transactions. The value and accuracy of the work, when estimated by standards which form proper criteria for that period, is universally recognised. It greatly promoted the progress of sidereal astronomy, and the authors were in consequence awarded medals from the Royal Society, and the Royal Astronomical Society, as well as similar testimonials from various foreign institutions.

This work must, however, be regarded as merely introductory to the main labours of John Herschel's life. His father devoted the greater part of his years as an observer to what he called his “sweeps “of the heavens. The great reflecting telescope, twenty feet long, was moved slowly up and down through an arc of about two degrees towards and from the pole, while the celestial panorama passed slowly in the course of the diurnal motion before the keenly watching eye of the astronomer. Whenever a double star traversed the field Herschel described it to his sister Caroline, who, as we have already mentioned, was his invariable assistant in his midnight-watches. When a nebula appeared, then he estimated its size and its brightness, he noticed whether it had a nucleus, or whether it had stars disposed in any significant manner with regard to it. He also dictated any other circumstance which he deemed worthy of record. These observations were duly committed to writing by the same faithful and indefatigable scribe, whose business it also was to take a memorandum of the exact position of the object as indicated by a dial placed in front of her desk, and connected with the telescope.

John Herschel undertook the important task of re-observing the various double stars and nebulae which had been discovered during these memorable vigils. The son, however, lacked one inestimable advantage which had been possessed by the father. John. Herschel had no assistant to discharge all those duties which Caroline had so efficiently accomplished. He had, therefore, to modify the system of sweeping previously adopted in order to enable all the work both of observing and of recording to be done by himself. This in many ways, was a great drawback to the work of the younger astronomer. The division of labour between the observer and the scribe enables a greatly increased quantity of work to be got through. It is also distinctly disadvantageous to an observer to have to use his eye at the telescope directly after he has been employing it for reading the graduations on a circle, by the light of a lamp, or for entering memoranda in a note book. Nebulae, especially, are often so excessively faint that they can only be properly observed by an eye which is in that highly sensitive condition which is obtained by long continuance in darkness. The frequent withdrawal of the eye from the dark field of the telescope, and the application of it to reading by artificial light, is very prejudicial to its use for the more delicate purpose. John Herschel, no doubt, availed himself of every precaution to mitigate the ill effects of this inconvenience as much as possible, but it must have told upon his labours as compared with those of his father.

But nevertheless John Herschel did great work during his "sweeps." He was specially particular to note all the double stars which presented themselves to his observation. Of course some little discretion must be allowed in deciding as to what degree of proximity in adjacent stars does actually bring them within the category of "double stars." Sir John set down all such objects as seemed to him likely to be of interest, and the results of his discoveries in this branch of astronomy amount to some thousands. Six or seven great memoirs in the Transactions of the Royal Astronomical Society have been devoted to giving an account of his labours in this department of astronomy.

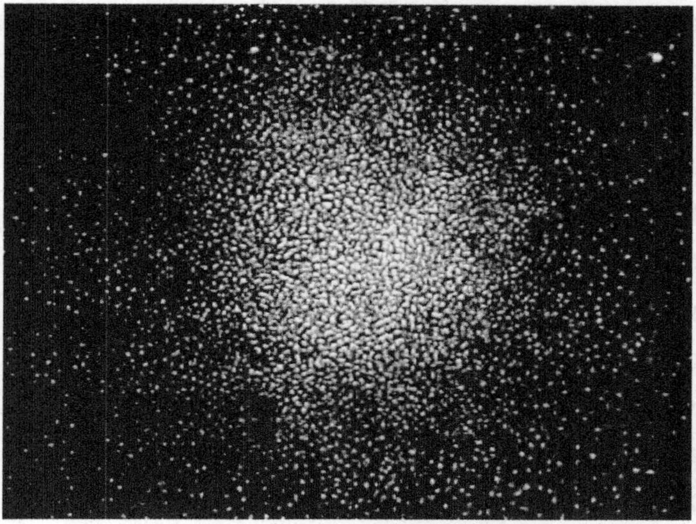

The cluster in the Centaur, drawn by Sir John Herschel.

One of the achievements by which Sir John Herschel is best known is his invention of a method by which the orbits of binary stars could be determined. It will be observed that when one star revolves around another in consequence of the law of gravitation, the orbit described must be an ellipse. This ellipse, however, generally speaking, appears to us more or less foreshortened, for it is easily seen that only under highly exceptional circumstances would the plane in which the stars move happen to be directly square to the line of view. It therefore follows that what we observe is not exactly the track of one star around the other; it is rather the projection of that track as seen on the surface of the sky. Now it is remarkable that this apparent path is still an ellipse. Herschel contrived a very ingenious and simple method by which he could discover from the observations the size and position of the ellipse in which the revolution actually takes place. He showed how, from the study of the apparent orbit of the star, and from certain measurements which could easily be effected upon it, the determination of the true ellipse in which the movement is performed could be arrived_ at. In other words, Herschel solved in a beautiful manner the problem of finding the true orbits of double stars. The importance of this work may be inferred from the fact that it has served as the basis on which scores of other investigators have studied the fascinating subject of the movement of binary stars.

The labours, both in the discovery and measurement of the double stars, and in the discussion of the observations with the object of finding the orbits of such stars as are in actual revolution, received due recognition in yet another gold medal awarded by the Royal Society. An address was delivered on the occasion by the Duke of Sussex (30th November, 1833), in the course of which, after stating that the medal had been conferred on Sir John Herschel, he remarks:—

“It has been said that distance of place confers the same privilege as distance of time, and I should gladly avail myself of the privilege which is thus afforded me by Sir John Herschel's separation from his country and friends, to express my admiration of his character in stronger terms than I should otherwise venture to use; for the language of panegyric, however sincerely it may flow from the heart, might be mistaken for that of flattery, if it could not thus claim somewhat of an historical character; but his great attainments in almost every department of human knowledge, his fine powers as a philosophical writer, his great services and his distinguished devotion to science, the high principles which have regulated his conduct in every relation of life, and, above all, his engaging modesty, which is the crown of all his other virtues, presenting such a model of an accomplished philosopher as can rarely be found beyond the regions of fiction, demand abler pens than mine to describe them in adequate terms, however much inclined I might feel to undertake the task."

The first few lines of the eulogium just quoted allude to Herschel's absence from England. This was not merely an episode of interest in the career of Herschel, it was the occasion of one of the greatest scientific expeditions in the whole history of astronomy.

Herschel had, as we have seen, undertaken a revision of his father's "sweeps" for new objects, in those skies which are visible from our latitudes in the northern hemisphere. He had well-nigh completed this task. Zone by zone the whole of the heavens which could be observed from Windsor had passed under his review. He had added hundreds to the list of nebulae discovered by his father. He had announced thousands of double stars. At last, however, the great survey was accomplished. The contents of the northern hemisphere, so far at least as they could be disclosed by his telescope of twenty feet focal length, had been revealed.

But Herschel felt that this mighty task had to be supplemented by another of almost equal proportions, before it could be said that the twenty-foot telescope had done its work, It was only the northern half of the celestial sphere which had been fully explored.

Sir John Herschel's Observatory at Feldhausen, Cape of Good Hope.

The southern half was almost virgin territory, for no other astronomer was possessed of a telescope of such power as those which the Herschels had used. It is true, of course, that as a certain margin of the southern hemisphere was visible from these latitudes, it had been more or less scrutinized by observers in northern skies. And the glimpses which had thus been obtained of the celestial objects in the southern sky, were such as to make an eager astronomer long for a closer acquaintance with the celestial wonders of the south. The most glorious object in the sidereal heavens, the Great Nebula in Orion, lies indeed in that southern hemisphere to which the younger Herschel's attention now became directed. It fortunately happens, however, for votaries of astronomy all the world over, that Nature has kindly placed her most astounding object, the great Nebula in Orion, in such a favoured position, near the equator, that from a considerable range of latitudes, both north and south, the wonders of the Nebula can be explored. There are grounds for thinking that the southern heavens contain noteworthy objects which, on the whole, are nearer to the solar system than are the noteworthy objects in the northern skies. The nearest star whose distance is known, Alpha Centauri, lies in the southern hemisphere, and so also does the most splendid cluster of stars.

Influenced by the desire to examine these objects, Sir John Herschel determined to take his great telescope to a station in the southern hemisphere, and thus complete his survey of the sidereal heavens. The latitude of the Cape of Good Hope is such that a suitable site could be there found for his purpose. The purity of the skies in South Africa promised to provide for the astronomer those clear nights which his delicate task of surveying the nebulae would require.

On November 13, 1833, Sir John Herschel, who had by this time received the honour of knighthood from William IV., sailed from Portsmouth for the Cape of Good Hope, taking with him his gigantic instruments. After a voyage of two months, which was considered to be a fair passage in those days, he landed in Table Bay, and having duly reconnoitred various localities, he decided to place his observatory at a place called Feldhausen, about six miles from Cape Town, near the base of the Table Mountain. A commodious residence was there available, and in it he settled with his family. A temporary building was erected to contain the equatorial, but the great twenty-foot telescope was accommodated with no more shelter than is provided by the open canopy of heaven.

As in his earlier researches at home, the attention of the great astronomer at the Cape of Good Hope was chiefly directed to the measurement of the relative positions and distances apart of the double stars, and to the close examination of the nebulae. In the delineation of the form of these latter objects Herschel found ample employment for his skilful pencil. Many of the drawings he has made of the celestial wonders in the southern sky are admirable examples of celestial portraiture.

The number of the nebulae and of those kindred objects, the star clusters, which Herschel studied in the southern heavens, during four years of delightful labour, amount in all to one thousand seven hundred and seven. His notes on their appearance, and the determinations of their positions, as well as his measurements of double stars, and much other valuable astronomical research, were published in a splendid volume, brought out at the cost of the Duke of Northumberland. This is, indeed, a monumental work, full of interesting and instructive reading for anyone who has a taste for astronomy.

Herschel had the good fortune to be at the Cape on the occasion of the periodical return of Halley's great comet in 1833. To the study of this body he gave assiduous attention, and the records of his observations form one of the most interesting chapters in that remarkable volume to which we have just referred.

Granite Column at Feldhausen, Cape Town, to commemorate Sir John Herschel's survey of the Southern Heavens.

Early in 1838 Sir John Herschel returned to England. Lie had made many friends at the Cape, who deeply sympathised with his self-imposed labours while he was resident among them. They desired to preserve the recollection of this visit, which would always, they considered, be a source of gratification in the colony. Accordingly, a number of scientific friends in that part of the world raised a monument with a suitable inscription, on the spot which had been occupied by the great twenty-foot reflector at Feldhausen.

His return to England after five years of absence was naturally an occasion for much rejoicing among the lovers of astronomy. He was entertained at a memorable banquet, and the Queen, at her coronation, made him a baronet. His famous aunt Caroline, at that time aged eighty, was still in the enjoyment of her faculties, and was able to estimate at its true value the further lustre which was added to the name she bore. But there is reason to believe that her satisfaction was not quite unmixed with other feelings. With whatever favour she might regard her nephew, he was still not the brother to whom her life had been devoted. So jealous was this vigorous old lady of the fame of the great brother William, that she could hardly hear with patience of the achievements of any other astronomer, and this failing existed in some degree even when that other astronomer happened to be her illustrious nephew.

With Sir John Herschel's survey of the Southern Hemisphere it may be said that his career as an observing astronomer came to a close. He did .not again engage in any systematic telescopic research. But it must not be inferred from this statement that he desisted from active astronomical work. It has been well observed that Sir John Herschel was perhaps the only astronomer who has studied with success, and advanced by original research, every department of the great science with which his name is associated. It was to some other branches of astronomy besides those concerned with looking through telescopes, that the rest of the astronomer's life was to be devoted.

To the general student Sir John Herschel is best known by the volume which he published under the title of "Outlines of Astronomy." This is, indeed, a masterly work, in which the characteristic difficulties of the subject are resolutely faced and expounded with as much simplicity as their nature will admit. As a literary effort this work is admirable, both on account of its picturesque language and the ennobling conceptions of the universe which it unfolds. The student who desires to become acquainted with those recondite departments of astronomy, in which the effects of the disturbing action of one planet upon the motions of another planet are considered, will turn to the chapters in Herschel's famous work on the subject. There he will find this complex matter elucidated, without resort to difficult mathematics. Edition after edition of this valuable work has appeared, and though the advances of modern astronomy have left somewhat out of date in certain departments, yet the expositions it contains of the fundamental parts of the science still remain unrivalled.

Another great work which Sir John undertook after his return from the Cape, was a natural climax to those labours on which his father and he had been occupied for so many years. We have already explained how the work of both these observers had been mainly devoted to the study of the nebulae and the star clusters. The results of their discoveries had been announced to the world in numerous isolated memoirs. The disjointed nature of these publications made their use very inconvenient. But still it was necessary for those who desired to study the marvellous objects discovered by the Herschels, to have frequent recourse to the original works. To incorporate all the several observations of nebulae into one great systematic catalogue, seemed, therefore, to be an indispensable condition of progress in this branch of knowledge. No one could have been so fitted for this task as Sir

John Herschel. He, therefore, attacked and carried through the great undertaking. Thus at last a grand catalogue of nebulae and clusters was produced. Never before was there so majestic an inventory. If we remember that each of the nebulae is an object so vast, that the whole of the solar system would form an inconsiderable speck by comparison, what are we to think of a collection in which these objects are enumerated in thousands? In this great catalogue we find arranged in systematic order all the nebula) and all the clusters which had been revealed by the diligence of the Herschels, father and son, in the Northern Hemisphere, and of the son alone in the Southern Hemisphere. Nor should we omit to mention that the labours of other astronomers were likewise incorporated. It was unavoidable that the descriptions given to each of the objects should be very slight. Abbreviations are used, which indicate that a nebula is bright, or very bright, or extremely bright, or faint, or very faint, or extremely faint.

Such phrases have certainly but a relative and technical meaning in such a catalogue. The nebulm entered as extremely bright by the experienced astronomer are only so described by way of contrast to the great majority of these delicate telescopic objects. Most of the nebulae, indeed, are so difficult to see, that they admit of but very slight description. It should be observed that Herschel's catalogue augmented the number of known nebulous objects to more than ten times that collected into any catalogue which had ever been compiled before the days of William Herschel's observing began. But the study of these objects still advances, and the great telescopes now in use could probably show at least twice as many of these objects as are contained in the list of Herschel, of which a new and enlarged edition has since been brought out by Dr. Dreyer.

One of the best illustrations of Sir John Herschel's literary powers is to be found in the address which he delivered at the

Royal Astronomical Society, on the occasion of presenting a medal to Mr. Francis Baily, in recognition of his catalogue of stars. The passage I shall here cite places in its proper aspect the true merit of the laborious duty involved in such a task as that which Mr, Baily had carried through with such success: —

“If we ask to what end magnificent establishments are maintained by states and sovereigns, furnished with masterpieces of art, and placed under the direction of men of first-rate talent and high-minded enthusiasm, sought out for those qualities among the foremost in the ranks of science, if we demand cui bono for what good a Bradley has toiled, or a Maskelyne or a Piazzi has worn out his venerable age in watching, the answer is— not to settle mere speculative points in the doctrine of the universe; not to cater for the pride of man by refined inquiries into the remoter mysteries of nature; not to trace the path of our system through space, or its history through past and future eternities. These, indeed, are noble ends and which I am far from any thought of depreciating; the mind swells in their contemplation, and attains in their pursuit an expansion and a hardihood which fit it for the boldest enterprise. But the direct practical utility of such labours is fully worthy of their speculative grandeur. The stars, are the landmarks of the universe; and, amidst the endless and complicated fluctuations of our system, seem placed by its Creator as guides and records, not merely to elevate our minds by the contemplation of what is vast, but to teach tis to direct our actions by reference to what is immutable in His works. It is, indeed, hardly possible to over appreciate their value in this point of view. Every well-determined star, from the moment its place is registered, becomes to the astronomer, the geographer, the navigator, the surveyor, a point of departure which can never deceive or fail him, the same forever and in all places, of a delicacy so

---

extreme as to be a test for every instrument yet invented by man, yet equally adapted for the most ordinary purposes; as available for regulating a town clock as for conducting a navy to the Indies; as effective for mapping down the intricacies of a petty barony as for adjusting the boundaries of Transatlantic empires. When once its place has been thoroughly ascertained and carefully recorded, the brazen circle with which that useful work was done may moulder, the marble pillar may totter on its base, and the astronomer himself survive only in the gratitude of posterity; but the record remains, and transfuses all its own exactness into every determination which takes it for a groundwork, giving to inferior instruments—nay, even to temporary contrivances, and to the observations of a few weeks or days—all the precision attained originally at the cost of so much time, labour, and expense."

Sir John Herschel wrote many other works

besides those we have mentioned. His "Treatise on Meteorology" is, indeed, a standard work on this subject, and numerous articles from the same pen on miscellaneous subjects, which have been collected and reprinted, seemed as a relaxation from his severe scientific studies. Like certain other great mathematicians Herschel was also a poet, and he published a translation of the Iliad into blank verse.

In his later years Sir John Herschel lived a retired life. For a brief period he had, indeed, been induced to accept the office of Master of the Mint. It was, however, evident that the routine of such an occupation was not in accordance with his tastes, and he gladly resigned it, to return to the seclusion of his study in his beautiful home at Collingwood, in Kent.

His health having gradually failed, he died on the 11th May, 1871, in the seventy-ninth year of his age.

- ROBERT STAWELL BALL

COMPLIMENTARY MATERIAL

A BRIEF BIOGRAPHICAL SKETCH

JOHN HERSCHEL

- ROBERT STAWELL BALL

John Frederick William **Herschel**
(March 7, 1792 – May 11, 1871)

COMPLIMENTARY MATERIAL

A BRIEF BIOGRAPHICAL SKETCH

JOHN HERSCHEL

Sir John Frederick William Herschel, 1st Baronet KH FRS (March 7, 1792 – May 11, 1871) was an English polymath, mathematician, astronomer, chemist, inventor, and experimental photographer, who also did valuable botanical work. He was the son of Mary Baldwin and astronomer William Herschel, nephew of astronomer Caroline Herschel and the father of twelve children.

Herschel originated the use of the Julian day system in astronomy. He named seven moons of Saturn and four moons of Uranus. He made many contributions to the science of photography, and investigated colour blindness and the chemical power of ultraviolet rays; his Preliminary Discourse (1831), which advocated an inductive approach to scientific experiment and theory building, was an important contribution to the philosophy of science.

Early Life and Work on Astronomy

Herschel was born in Slough, Buckinghamshire, the son of Mary Baldwin and William Herschel. He studied shortly at Eton College and St John's College, Cambridge, graduating as Senior Wrangler in 1813. It was during his time as an undergraduate that he became friends with Charles Babbage and George Peacock. He left Cambridge in 1816 and started working with his father. He took up astronomy in 1816, building a reflecting telescope with a mirror 18 inches (460 mm) in diameter and with a 20-foot (6.1 m) focal length. Between 1821 and 1823 he re-examined, with James South, the double stars catalogued by his father. He was one of the founders of the Royal Astronomical Society in 1820. For his work with his father, he was presented with the Gold Medal of the Royal Astronomical Society in 1826 (which he won again in 1836), and with the Lalande Medal of the French Academy of Sciences in 1825, while in 1821 the Royal Society bestowed upon him the Copley Medal for his mathematical contributions to their Transactions. Herschel was made a Knight of the Royal Guelphic Order in 1831.

He served as President of the Royal Astronomical Society three times: 1827–29, 1839–41 and 1847–49.

His A *preliminary discourse on the study of natural philosophy*, published early in 1831 as part of Dionysius Lardner's Cabinet cyclopædia, set out methods of scientific investigation with an orderly relationship between observation and theorising. He described nature as being governed by laws which were difficult to discern or to state mathematically, and the highest aim of natural philosophy was understanding these laws through inductive reasoning, finding a single unifying explanation for a phenomenon. This became an authoritative statement with wide influence on science, particularly at the University of Cambridge where it inspired the student Charles Darwin with "a burning zeal" to contribute to this work.

Herschel published a catalogue of his astronomical observations in 1864, as the *General Catalogue of Nebulae and Clusters,* a compilation of his own work and that of his father's, expanding on the senior Hershel's *Catalogue of Nebulae.* A further complementary volume was published posthumously, as the *General Catalogue of 10,300 Multiple and Double Stars.*

Calotype (of model) of lunar crater Copernicus, 1842

Herschel correctly considered astigmatism to be due to irregularity of the cornea and theorised that vision could be improved by the application of some animal jelly contained in a capsule of glass against the cornea. His views were published in an article entitled Light in 1828 and the Encyclopædia Metropolitana in 1845.

Visit to South Africa

Declining an offer from the Duke of Sussex that they travel to South Africa on a Navy ship, Herschel and his wife paid £500 for passage on the S.S. Mountstuart Elphinstone, which departed from Portsmouth on 13 November 1833. The voyage to South Africa was made in order to catalogue the stars, nebulae, and other objects of the southern skies. This was to be a completion as well as extension of the survey of the northern heavens undertaken initially by his father William Herschel. He arrived in Cape Town on 15 January 1834 and set up a private 21 ft (6.4 m) telescope at Feldhausen at Claremont, a suburb of Cape Town. Amongst his other observations during this time was that of the return of Comet Halley. Herschel collaborated with Thomas Maclear, the Astronomer Royal at the Cape of Good Hope and the members of the two families became close friends. During this time, he also witnessed the Great Eruption of Eta Carinae (December, 1837).

- ROBERT STAWELL BALL

John Herschel, 1846

In addition to his astronomical work, however, this voyage to a far corner of the British empire also gave Herschel an escape from the pressures under which he found himself in London, where he was one of the most sought-after of all British men of science. While in southern Africa, he engaged in a broad variety of scientific pursuits free from a sense of strong obligations to a larger scientific community. It was, he later recalled, probably the happiest time in his life.

In an extraordinary departure from astronomy, he combined his talents with those of his wife, Margaret, and between 1834 and 1838 they produced 131 botanical illustrations of fine quality, showing the Cape flora. Herschel used a camera lucida to obtain accurate outlines of the specimens and left the details to his wife. Even though their portfolio had been intended as a personal record, and despite the lack of floral dissections in the paintings, their accurate rendition makes them more valuable than many contemporary collections. Some 112 of the 132 known flower studies were collected and published as Flora Herscheliana in 1996.

As their home during their stay in the Cape, the Herschels had selected 'Feldhausen' ("Field Houses"), an old estate on the south-eastern side of Table Mountain. Here John set up his reflector to begin his survey of the southern skies. Herschel, meanwhile, read widely. Intrigued by the ideas of gradual formation of landscapes set out in Charles Lyell's Principles of Geology, he wrote to Lyell on 20 February 1836 praising the book as a work that would bring "a complete revolution in [its] subject, by altering entirely the point of view in which it must thenceforward be contemplated" and opening a way for bold speculation on "that mystery of mysteries, the replacement of extinct species by others." Herschel himself thought catastrophic extinction and renewal "an inadequate conception of the Creator" and by analogy with other intermediate causes, "the origination of fresh species, could it ever come under our cognizance, would be found to be a natural in contradistinction to a miraculous process". He prefaced his words with the couplet:

He that on such quest would go must know not fear or failing

To coward soul or faithless heart the search were unavailing.

Taking a gradualist view of development and referring to the evolution of language, he commented:

Words are to the Anthropologist what rolled pebbles are to the Geologist — battered relics of past ages often containing within them indelible records capable of intelligent interpretation — and when we see what amount of change 2000 years has been able to produce in the languages of Greece & Italy or 1000 in those of Germany France & Spain we naturally begin to ask how long a period must have lapsed since the Chinese, the Hebrew, the Delaware & the Malesass [Malagasy] had a point in common with the German & Italian & each other — Time! Time! Time! — we must not impugn the Scripture Chronology, but we must interpret it in accordance with whatever shall appear on fair enquiry to be the truth for there cannot be two truths. And really there is scope enough: for the lives of the Patriarchs may as reasonably be extended to 5000 or 50000 years apiece as the days of Creation to as many thousand millions of years.

The document was circulated, and Charles Babbage incorporated extracts in his ninth and unofficial *Bridgewater Treatise*, which postulated laws set up by a divine programmer. When HMS Beagle called at Cape Town, Captain Robert FitzRoy and the young naturalist Charles Darwin visited Herschel on 3 June 1836. Later on, Darwin would be influenced by Herschel's writings in developing his theory advanced in The Origin of Species. In the opening lines of that work, Darwin writes that his intent is "to throw some light on the origin of species — that mystery of mysteries, as it has been called by one of our greatest philosophers," referring to Herschel.

Herschel returned to England in 1838, was

created a baronet, of Slough in the County of Buckingham, and published Results of Astronomical Observations made at the Cape of Good Hope in 1847. In this publication he proposed the names still used today for the seven then-known satellites of Saturn: Mimas, Enceladus, Tethys, Dione, Rhea, Titan, and Iapetus. In the same year, Herschel received his second Copley Medal from the Royal Society for this work. A few years later, in 1852, he proposed the names still used today for the four then-known satellites of Uranus: Ariel, Umbriel, Titania, and Oberon.
Portrait of Sir John Herschel by his daughter Margaret Louisa Herschel

Photography

Herschel made numerous important contributions to photography. He made improvements in photographic processes, particularly in inventing the cyanotype process and variations (such as the chrysotype), the precursors of the modern blueprint process. In 1839, he made a photograph on glass, which still exists, and experimented with some color reproduction, noting that rays of different parts of the spectrum tended to impart their own color to a photographic paper. Herschel made experiments using photosensitive emulsions of vegetable juices, called phytotypes and published his discoveries in the Philosophical Transactions of the Royal Society of London in 1842. He collaborated in the early 1840s with Henry Collen, portrait painter to Queen Victoria.

Herschel originally discovered the platinum process on the basis of the light sensitivity of platinum salts, later developed by William Willis.

- ROBERT STAWELL BALL

Portrait of **Sir John Herschel** by his daughter
Margaret Louisa Herschel

Herschel coined the term photography in 1839. He may however have been preceded by Brazilian Hércules Florence, who used the French equivalent, photographie, to describe his own experiments in private notes in 1834, although as Florence was not in communication with the European scientific community, Herschel has historically been credited with coining and popularising the term. Herschel was also the first to apply the terms negative and positive to photography.

He discovered sodium thiosulfate to be a solvent of silver halides in 1819, and informed Talbot and Daguerre of his discovery that this "hyposulphite of soda" ("hypo") could be used as a photographic fixer, to "fix" pictures and make them permanent, after experimentally applying it thus in early 1839. His ground-breaking research on the subject was read at the Royal Society in London in March 1839 and January 1840.

General

Tombs of John Herschel and Charles Darwin. Westminster Abbey.
Description of a machine for resolving by inspection certain important forms, 1832

Herschel wrote many papers and articles, including entries on meteorology, physical geography and the telescope for the eighth edition of the Encyclopædia Britannica. He also translated the Iliad of Homer.

He invented the actinometer in 1825 to measure the direct heating power of the sun's rays, and his work with the instrument is of great importance in the early history of photochemistry.

He proposed a correction to the Gregorian calendar, making years that are multiples of 4000 not leap years, thus reducing the average length of the calendar year from 365.2425 days to 365.24225.[24] Although this is closer to the mean tropical year of 365.24219 days, his proposal has never been adopted because the Gregorian calendar is based on the mean time between vernal equinoxes (currently 365.242374 days).

Hershel was elected a Foreign Honorary Member of the American Academy of Arts and Sciences in 1832, and in 1836, a foreign member of the Royal Swedish Academy of Sciences.

In 1835, the New York Sun newspaper wrote a series of satiric articles that came to be known as the Great Moon Hoax, with statements falsely attributed to Herschel about his supposed discoveries of animals living on the Moon, including batlike winged humanoids.

The village of Herschel in western Saskatchewan (Canada), Mount Herschel (Antarctica), the crater J. Herschel on the Moon, and the Herschel Girls' School in Cape Town (South Africa), are all named after him. While it is commonly accepted that Herschel Island (in the Arctic Ocean, part of the Yukon Territory) was named after him, the entries in the expedition journal of Sir John Franklin state that the latter wished to honour the Herschel name, about which John Herschel's father (Sir William Herschel) and his aunt (Caroline Herschel) are two other notable members of this family.

Family

John Frederick William Herschel
by Alfred Edward Chalon, 1829.

Margaret Brodie Stewart
by Alfred Edward Chalon, 1829.

He married his cousin Margaret Brodie Stewart (1810–1884) on 3 March 1829 at Edinburgh and was father of the following children:

1. Caroline Emilia Elizabeth Herschel (31 March 1830 – 29 Jan 1909), who married Alexander Hamilton-Gordon

2. Isabella Herschel (5 June 1831 – 1893)

3. Sir William James Herschel, 2nd Bt. (9 January 1833 – 1917),

4. Margaret Louisa Herschel (1834–1861), an accomplished artist

5. Prof. Alexander Stewart Herschel (1836–1907), FRS

6. Col. John Herschel FRS, FRAS, (1837–1921) surveyor

7. Marie Sophie Herschel (1839–1929)

8. Amelia Herschel (1841–1926) married Sir Thomas Francis Wade, diplomat and sinologist

9. Julia Edith Herschel (1842–1933) married on 4 June 1878 to Captain (later Admiral) John Fiot Lee Pearse Maclear

---

- ROBERT STAWELL BALL

10. Matilda Rose Herschel (1844–1914)

11. Francisca Herschel (1846–1932)

12. Constance Ann Herschel (1855–20 June 1939)

On his death at Collingwood, his home near Hawkhurst in Kent, he was given a national funeral and buried in Westminster Abbey.

Tombs of John Herschel and Charles Darwin. Westminster Abbey.

Bibliography

- *On the Aberration of Compound Lenses and Object-Glasses (1821)*
- Book-length articles on "Light", "Sound" and "Physical Astronomy" for the *Encyclopaedia Metropolitana* (30 vols. 1817-45)
- *A preliminary discourse on the study of natural philosophy, part of Dionysius Lardner's Cabinet cyclopœdia (1831, new edition 1840)*
- *A Treatise on Astronomy (1833)*
- *Outlines of Astronomy (1849)*
- *General Catalogue of 10,300 Multiple and Double Stars (published posthumously)*
- *Familiar Lectures on Scientific Subjects*

- *General Catalogue of Nebulae and Clusters*
- *Manual of Scientific Inquiry (ed.), (1849)*
- *Meteorology (1861)*
- *Physical Geography (1861)*
- *Familiar Lectures on Scientific Subjects (1867)*

- Herschel, John F. W. (1840-02-20), "On the Chemical Action of the Rays of the Solar Spectrum on Preparations of Silver and Other Substances, Both Metallic and Non-Metallic, and on Some Photographic Processes", Philosophical Transactions of the Royal Society of London, 130: 1–59, Bibcode:
1840RSPT..130....1H,
doi:10.1098/rstl.1840.0002,
ISSN 0261-0523, JSTOR 108209.

ABOUT THE AUTHOR

ROBERT STAWELL BALL

(July 1, 1840 – November 25, 1913)

Sir Robert Stawell Ball was an Irish astronomer who founded the screw theory.

He was the son of naturalist Robert Ball and Amelia Gresley Hellicar. He was born in Dublin.

Ball worked for Lord Rosse from 1865 to 1867. In 1867 he became Professor of Applied Mathematics at the Royal College of Science in Dublin. There he lectured on mechanics and published an elementary account of the science.

In 1874 Ball was appointed Royal Astronomer of Ireland and Andrews Professor of Astronomy in the University of Dublin at Dunsink Observatory.

Ball contributed to the science of kinematics by delineating the screw displacement:

When Ball and the screw theorists speak of screws they no longer mean actual cylindrical objects with helical threads cut into them but the possible motion of any body whatsoever, including that of the screw independently of the nut.

Ball's treatise *The Theory of Screws* (1876) is now in the public domain. His work on screw dynamics earned him in 1879 the

Cunningham Medal of the Royal Irish Academy.

In 1882 Popular Science Monthly carried his article "*A Glimpse through the Corridors of Time*". The following year it carried his two-part article on "*The Boundaries of Astronomy*".

Ball expounded the tides in *Time and Tide*: a Romance of the Moon In 1892 he was appointed Lowndean Professor of Astronomy and Geometry at Cambridge University at the same time becoming director of the Cambridge Observatory. He was a fellow of King's College, Cambridge.

In 1900 Cambridge University Press published *A Treatise on the Theory of Screws.* That year he also published *The Story of the Heavens.* Much in the limelight, he stood as President of the Quaternion Society. He was also President of the Mathematical Association in 1900.

In 1908 he published *A Treatise on Spherical Astronomy*, which is a textbook on astronomy starting from spherical trigonometry and the celestial sphere, considering atmospheric refraction and aberration of light, and introducing basic use of a generalised instrument.

His work *The Story of the Heavens* is mentioned in the "Ithaka" chapter of Ulysses. His lectures, articles and books (e.g. *Starland* and *The Story of the Heavens*) were mostly popular and simple in style.

He died in Cambridge and was buried at the Parish of the Ascension Burial Ground in Cambridge, with his wife Lady Francis Elizabeth Ball. Their children were: Frances Amelia, Robert Steele, William Valentine (later Sir), Mary Agnetta, Charles Rowan Hamilton, and Randall Gresley (later Colonel).

LECTURES

Ball became celebrated for his popular lectures on science. He gave an estimated 2500 lectures between 1875 and 1910 in towns and cities across Britain and Ireland. In 1892, 1898 and 1900 he was invited to deliver the Royal Institution Christmas Lecture. *Astronomy; Astronomy* and *Great Chapters from the Book of Nature.*

FEEDBACK

Now that you have read the book ...

Was it interesting?

Did you enjoy what you wanted to read?
Was there any room for improvement?

Let us know at:

http://www.diamondbooks.ca/feedback

Your feedback is highly appreciated.
Thank you!

Made in the USA
Coppell, TX
22 January 2022

72118622R00066